STEELMILL BLUES

Steve Packard

Singlejack Books
San Pedro 1978

ISBN: 0-917300-03-3
Manufactured in the U.S.A.

28

The publisher wishes to thank
The Liberty hill Foundation, Los Angeles,
for their grant
which made this first edition possible.

The term *singlejack* originated with the hardrock miners of the American west. The drilling of holes for the insertion of dynamite was a tough and dangerous job. The miners worked in pairs, with one kneeling to hold erect the steel drill, which he would turn slowly as his partner drove it into the rock with blows from a sledge (or single jack) hammer. They would switch tasks now and then, and since the job demanded as much mutual trust as skill, many lasting friendships were formed.

Around the turn of the Century, on-the-job organizers for the Western Federation of Miners and the Industrial Workers of the World brought additional meaning to *singlejack*. They used it to describe that method of organizing where dedicated advocates are developed one at a time on a highly personalized basis - as between partners.

In turn, still broader veins of definition can be extracted from this rich historical term. We like to apply it to that private bond which ideally is sparked between a reader and a book. We hope you agree.

Miles & Weir, Ltd.
Singlejack Books
Box 1906
San Pedro, California 90733

Thanks

to Staughton Lynd, who was the first person to treat these stories seriously — which made all the difference in the world to me.

U.S. Steel. Big Deal. Largest steel mill in the country — Gary, Indiana. I could of got other work, if I wasn't so rebellious, pushing a pencil in some office no doubt. More money probably, and ulcers sooner or later.

I want a job because I'm broke, frankly, but also because I need some sort of new direction. The steelworkers are supposed to strike six months from now. People say the Gary Works may become a pretty wild place, and some good changes may be made then.

Fire, steam, hulking gray structures, the mill floats around me. Like when you start at any new place, so many things are happening at the same time you feel a little stoned. Probably you can steer your way through, but only vaguely in touch with half of what passes by.

My division is called Merchant Mills. The

basic warehouse building is just a couple of blocks wide. But it's so long if you stand in the middle you can't see halfway through the gloom to either end. Huge doors are everywhere, always open; it's February and snow piles in around the edges.

On the east side of the place are the twelve actual "mills," sticking out from the main building like the teeth of an Afro comb. Each of these long skinny rolling mills is the length of a football field. Red-hot ribbons of steel come snaking streaking down them. By the time the ribbons get to us they're usually gray, but still hot enough to cook your lunch on. Our job is to cut steel to the right lengths. To test and inspect various shapes and kinds for special uses. To straighten or flatten them in enormous noisy machines. To finally tie piles with steel wire, or bind them with metal bands.

Every morning I get to Gary about fifteen minutes before work. The whole city-sized mill complex is surrounded by a high metal fence and a dark moat of chemical-smelling water, once a river, now the lower end of the alimentary canal that feeds U.S. Steel's waste products into Lake Michigan. The inner defense perimeter consists of a guardhouse and a gate where armed men look at our I.D.

cards, make sure we match our pictures, then give us our daily I.B.M. cards. I write my name and company number on the top of it.

Next, to the gray washhouse, and changing into filthy clothes. After one day they're the kind of dirty that lets them stand up in the locker by themselves. Regulation metatarsal shoes with metal plates over the top so it's harder to break your foot. I pull on a couple of sweatshirts and an army coat. Walking toward Merchant Mills I gather the long hair off my shoulders, twist it around a few times, and trap it under my hard hat where it stays all day. Finally the safety glasses — little wire screens fold back against your temples — special adjustable rims curl securely around your ears and hurt in every possible position.

The foreman's name is Chuck. He looks us over, who's here today and who's not. Then he mentally figures everyone's seniority against the jobs that need doing. He makes notes on his clipboard, and reads off the orders. "Packard, Quintana, Lewis, Pritchet: coil trimmer!" We give him our I.B.M. cards, and he makes a note on each, since different jobs get different pay. Most of the jobs I get rate $2.81 per hour for day shift and $2.89 at night.

As we each hand him our I.B.M. card, the foreman reads aloud the top "safety check" card on his deck and then puts it on the bottom, like the Community Chest pile in a Monopoly game: "Watch for slippery grease spots on floor..." "Wear your hard hat at all times..." Makes you feel a bit stupid and a bit rebellious at the start of every day.

Forty years ago, workers fought to unionize these mills. Gary has a proud and powerful history, where the army was called out many times to suppress militant workers and prevent the establishment of a union. But people fought on and sacrificed for many years and the union finally won.

Or at least that's what I read in radical history books.

No one from the union has ever talked to me. I've never been issued a "union card." (What would I do with it if I had one, I wonder.) I never see a sign, a poster, a leaflet, or announcement, passed out or posted up, by the union. People don't discuss the union. I wonder, does it have meetings? Do they vote? Is there a union hall? And all the heroism and bloodshed, the years of hard organizing work, what were they for?

Actually, I was afraid the union would be

racist and conservative, and we would have to struggle against them. I didn't expect that there would just be no union. I ask people if they know who our grievance man is, and almost nobody does. The union takes a few bucks out of my paycheck; I can see it listed over by the taxes and the F.I.C.A. and all the other paycheck leaks. That's my only evidence it still exists.

Is this mill just unorganized chaos, basically a scene of individual workers against the company? Now I've been here a while, and I'm slowly learning. Gary Works is a very complex organism, thousands of people and machines making up a living, changing thing. It's not organized too clearly by the company, not organized by the union at all. But very organized, just the same. I got my first sense of it from this story, told by my friend Spider, a black craneman who I often ride to work with.

One day, a white craneman was assigned to a good crane that should have gone to a certain black. The foreman made some little excuse to explain it, but really he was doing a favor for a white friend, and at the same time showing everyone who's boss. This foreman was new or he never would have been so foolish.

Black cranemen decided to sabotage production until this bullshit was straightened out. They had mild support from most white cranemen, who also thought the foreman was wrong.

Nothing can operate long without the cranes bringing and taking steel, so blacks quietly stopped the whole mill. They kept the cranes in lowest gear and worked in super-slow-motion. Foremen soon began hatching out of their offices, looking around, rubbing their eyes in disbelief. It was like the whole building popped LSD or the air had turned to some thick jelly: everything but the foremen moved at one-tenth normal speed.

Steel was piling up, and men stopping work. Foremen yelling, and cranemen explaining all the complex problems their machines had somehow suddenly developed. Calls were coming in from the main office in Pittsburgh, where the computer showed that our statistics had all stopped. So of course the company quick called the union.

Union sharpies swooped down from nowhere. "Men. Men. Come on now. What is the problem? Let's go through the channels on this thing. Right? Submit a grievance. What you're doing is an illegal strike!" But everyone knew that grievances take months, and that the day this foreman had to be wised-up was today.

After a couple of hours the company backed down. Higher level bosses appeared and announced that the foreman had "made a mistake." They switched around the black and white cranemen and the slowness thawed right out and the mill began to hum again at its normal rate. According to the rules. The worker's rules.

Big wildcats like that are less common than small, partial ones. The most often used weapon is sabotage. From the wholesale wrecking of machines to trickier, more selective actions — endless sabotages go on every day. I'm on many jobs where the foreman tries to hustle us too much or insists that we do a job in some dangerous or unpleasant way, and somebody breaks a machine. Nothing can be proved: only the workers know just what combination of speed, pressure, change of pace, angle of feed, is sufficient to make some dirty old monster start to shake and throw sparks, and smoke to a crunching halt. How fast the millwright fixes it depends on what vibrations he feels from the waiting men. And of course the foreman swears, and fumes, and groans, but he can't really do anything. If he is not overly dumb, he learns the limits beyond which he cannot go.

The men don't usually talk about this stuff; communication is carried out through undercurrents and understandings. But everyone in the mill agrees that the workers are right, which is why there is nothing the boss can do. We know the ins and outs of the machines enough to use them, undetected, against the boss, whenever necessary.

The only way a foreman can survive — the only way he gets a fair amount of work done in his zone — is to understand this communication-by-sabotage. It is the workers' police and court of law, by which we struggle to enforce our code of what's right and fair. It is part of a system laboriously built up over the decades.

I feel like I'm doing well, so far. I've played good roles in a number of little rebellions and have growing mutual respect and good feelings with guys in all parts of our mill. This process is a tricky one though, because the place is strongly divided up into minority groups. At line-up every morning all the workers are either black, or Latin, or hillbilly, or old ethnic European, or young white hippy-type. I'm hippy-type.

Nearly every foreman is "middle-America"

white. Other older whites have skilled jobs and don't line up with our "labor gang." The workers most mistrusted by the other minorities are hillbillies and hippies — because we occasionally get transformed into bosses.

Yesterday, Hendrickson, one of the foreman's sidekicks and henchmen, drew me aside and gave me a little pep talk about a new job he wants me to learn: the job of "inspector."

Hendrickson explains that "it's a real smart job to learn, Steve, and I think you can do it. The work's not hard. You don't have much time (seniority), that's true. But people here like you; I don't think anyone will complain. And it's a good way to get started out of this labor gang." What he means is that young whites should want to get ahead, and this inspector's job is a first step toward having my head fitted for a foreman's blue hard hat.

Today Chuck, the foreman, sends my friend Chico and me to the inspection tables, together. Hendrickson takes me aside and says, "Look, Steve, here's your clipboard, and I'm going to show you how to do this job. Give it your best, son, O. K.? We really could use another inspector. Most of these guys, you know..." He looks around to see that no non-white is listening, and then says, "Most of these guys...fucking Puerto Rican and the colored...don't even know how to read and write."

The picture becomes even clearer. Hendrickson shows me a couple of simple-minded things to do, with a micrometer, a pencil, and that clipboard. Then he says, "Sanchez here is your helper," pointing to my friend Chico, who is scowling. "Your job, Packard, is to tell him when to move steel for you. Have him spread out each bundle of rods on this bed. Then you check them, and have Sanchez pull them down into that trough and wire them up again. You inspect. That's all. You don't do this other work. Understand?"

Once or twice the thought flashes through my mind, why not become an inspector? But it's just not me. I decide to do the job badly, so I won't be asked again, and Chico is a big help in this. He won't work.

I say, "What a drag. I can't dig all this 'inspecting' bullshit. But I got to finish out the day. Let's work slow." Chico doesn't go for this line. No matter what I do, or say, I am neck deep in hostility. Chico doesn't lift a finger. He says, "I don't think you make a good inspector, hippy." The other two Spanish guys always call me hippy, but Chico never has before.

He moves slow, uncooperative, messes

things up on purpose. Most of the time he's in the washhouse or the canteen or sitting around half-asleep. He teases me, wants me to break down and order him to work. Chico was my friend, but now he's baiting me — waiting, almost wanting me to use that angry pissed-off commanding voice. If I say, "For Christ sakes, Chico, we got to do some work today. Let's go!" he'll start working real fast and be my enemy forever.

Twice they catch me doing the job Chico is supposed to do: "Oh. Packard! I guess you didn't understand. You don't move the steel, you have your helper do that. You inspect, all right? Have Sanchez do the rest." Sure.

I don't push Chico. We get almost nothing done, and Chuck is freaked. The next day he puts me back on inspecting, but with a new helper, a black man about forty years old. For some reason this guy thinks he's supposed to work hard. Has he just arrived from the South? I can't talk to him very well.

We do five or six times more steel than I did with Chico, but I save my skin by rejecting two-thirds of it. The reject pile grows bigger than I've ever seen it. Chuck comes, furious again. With him is the chief inspector, who's friendly and fatherly. The chief looks over my steel and admits that it's not quite up to the specifications in the book, but sort of winks and says that we might stretch the limits just a bit (wink) and accept most of this stuff.

I say thanks. And continue to reject two out of three loads. I think of the various mortar shells or peasant-village-defoliation-sprayers this steel might be destined to make. I think how much I hate being Mr. Inspector. The cranes hoist load after shiny load back to the scrap cars.

Days come and days go. I never get an upwardly mobile job again.

One of my favorite guys to work with is Billy. Billy will be a skilled worker one day, but he's in no hurry. He's generous, hates privilege, learned a lot of politics in Vietnam, always sides with the minorities against the foreman.

To watch him work is to watch a performance: he's a master. Billy runs his gray hulking band-leveler like he's playing on an organ. He knows every possible subtle adjustment, combination of lubricants, points of strain, and everything else necessary to play this massive thing like no one else alive.

He knows what effect X, Y, or Z lubrication in the front bearings will have on the permissible tension of angle lever way down on the other end (and so, of course, he uses these to magnificently cripple the mother whenever Chuck steps out of line. Which isn't often).

Another area of brilliance are the contraptions Billy devises out of Coke cans, scrap wood, cardboard, etc.. Once he rigged up a plastic milk carton and a long bent nail to do 95% of the job I was assigned that day (this automation was removed when Chuck lurked about, of course). If you asked Billy, he'd tell you he constructs these inventions to make his work easier, but it's clear to me he does them for the sheer joy and challenge.

Billy doesn't produce a lot of steel. He quickly admits he could get out two or three more times as much with little sweat. But every experienced man who works this machine turns out the same small amount. It's part of our speed-up war with the company, and it would betray the other workers if you did more.

So my friend uses his energy to teach new people tricks, to build a fire in the salamander for the crew on the next shift, to perform, and to clean his machine with thorough tenderness, scrub and oil it like a prized sports car, or a precious baby, for a good hour at the end of every day.

I would like to understand Billy. He talks reluctantly about his bored television-watching marriage. Less reluctantly about the nutty things he spends his money on. And then there are virtual hymns of devotion to Lake Michigan trout and to his boat. She's the only thing in his life he really speaks of fondly, that boat, and he worked a hell of a lot of overtime to get her. (Though now Billy works overtime to buy a bigger one.) I have a feeling, from Billy, and from many people here, that the things they buy with their money are often more excuses to spend, than real needs or wants. I think Billy doesn't know what to do with himself. I think he spends money so he'll have to work. I begin to toy with a new slogan: "Work is the opiate of the people."

More and more of my day becomes deep thoughts and ramblings. When I came here I had vague, radical ideas abut being an organizer. Lately I am just an observer. A lot of young guys cope with the emptiness here by dope or pills. But it's the middle-agers and the older guys I feel the worst for.

Scattered here and there in the gloom are things that look like heavy-steel four-legged garbage-cans, punched with holes and packed with burning coal to provide a little heat. For some reason they're called "salamanders." Circles of men sit on kegs and benches around the salamanders for three or four hours a day. Sometimes these adults talk, but mostly they just sit there with their lights out. As if they're watching TV, but with no TV.

The foremen can't make you work harder. They've given up trying. But they won't let you do anything else. You can't play music, you can't fool around, if they catch you reading you get yelled at. Sometimes I use my time to write all these ideas down, and I always put "Dear Carol" at the top of the page so people won't ask questions. But I can't let a foreman catch me writing. The only thing they let you do here is sit, like a vegetable.

Once a foreman came by a salamander where seven or eight of us were growing out of a couple of benches, one of us snoozing. The foreman kicked the guy's foot, shook him and grumbled, "What's this? A sleep-in? Look a little bit alive here!!" So the guy opened his eyes and sat up straight. He looked a little bit alive. The foreman was satisfied — and walked slowly on away through the mill. We chuckled.

One day, toward the tail end of winter, I

finish my job in an hour, and go wandering off across this mammoth complex, through U.S. Steel's forbidden "Inner City," slowly toward the shore of Lake Michigan. I walk through enormous railroad yards, wide weedy prairies struck with room-sized blocks of black rusting solid steel. It seems like the surface of a strange planet. Or the burial ground of an ancient civilization. I stop to watch the work in the wheel mill (where train wheels are made); I prowl around the plate mill where Ping-Pong-table sized ponds of molten steel burn out silently of heat and color. I watch the technology and poetry and spectacle of these places, I think of how my spirit would glow with pride at this incredible mass of human accomplishment if it was all being done for the good.

At the cold water's edge, wind and spray have piled crude icebergs bigger than the steel blocks behind me. Crashing waves, white and gray and winter blue-black, bright sun, purple sky with intense red-orange clouds from the mill's smokestacks. The water is filthy, and the icebergs speckled with oil and cans and boxes and crap, but if you squint your eyes they are towering white and beautiful. When I get back to Merchant Mills it's two and a half hours after I left. No one has missed me. I'm bored.

Next day I begin my time-and-motion study, a little bit of "social science" like I learned in college. I walk up and down the aisles noting on a piece of paper who's doing what, being careful that no one sees me and decides I'm a company spy.

In a half-hour I collect the following data:

Men apparently working: 80

Men walking around (some may have been doing a job, but most were just killing time like me): 29

Men sitting around salamanders: 64

Men standing around doing nothing: 53

Men dancing: 1

Over half are wasting their time away. I guess I knew it would be this bad. They think they're getting away with something. I have no crocodile tears about how the war effort won't have enough steel, but these numbers are about the emptiness of people's lives.

I make my survey again, picking times when it is most likely people will be working if they ever work. The last day of this I have plenty of time, because my job for eight hours is to sweep a little dirt, steel filings, and cigarette packages out of an area the size of a living room. Here's the total:

People working: 366

People doing nothing: 600

Someone's been writing chalk slogans on the wall of the wash-house: "Keep the faith, baby." Work ½ out of eight hours and you will KEEP THE FAITH." "Be a soul man, not a Co. man... Work ½ out of eight hours." They are all signed by "Cadillac Louis." A general feeling among all but the hillbilly workers is that it's right to sluff off. And there is something revolutionary about the workers' controlling the pace of their work this way. Besides, if they didn't, there would be a lot more hungry unemployed around the country.

But there is something deadening and enslaving here, too. I look around the circles of soot-black hulking shapes sitting joyless, motionless, around the salamanders. They are like the molted skins of insects left by the shore of a pond. As if their souls have left them.

It's a strange atmosphere the coming strike makes. No one talks about it, no concern or excitement. We have been producing huge stockpiles of steel so that our strike won't hurt the company, even if we stay out for months. Everyone knows, and no one cares.

In the last week the company has already

started laying off men. There is so much steel now no one knows what to do with it, and half the men hired after me are already gone. We throw good-bye parties every afternoon, starting about a half-hour before quitting time. We take turns bringing the treats. One day I chip in a watermelon, next day Charles brings little cakes and ice cream, next day it's Otis with spare ribs and Cokes. Chuck would stop us if he could, but everyone is in a weird frame of mind these days, and he doesn't dare get tough. Some of the oldest men come over and ask questions about what we're doing. We explain, and they put mixed expressions on their faces, thinking about the days when a layoff or strike was a serious thing, and how times have changed.

I only learn about the strike vote meeting from a radical friend who works in the coke ovens. No one else in Merchant Mills seems to know a thing, so I call up the union, and a voice says impatiently, "Yes, the strike vote is Tuesday. Everybody's been informed." I tell her that not a damn person in Merchant Mills knows about it, and she promises that the union will look into this. That night I see there's one dim leaflet tacked up in the glass case by the foreman's office. I try to take the meeting up; but nobody cares. One older black guy sums up the feeling pretty well: "What difference? Meeting, no meeting. Strike, no strike. They going to do it just like they want anyhow." By "they" he means the union and the company.

Tuesday. The union hall of Local 1014 of the United Steelworkers of America is packed by the time I arrive, and I have to sit in the front. (I'm told later that my picture is on the cover of the union newspaper. But I've never seen our union newspaper.) About 400 people are at the meeting, out of 15,000 members.

Harry Piasecki, the union president, starts by announcing in a tough and determined voice that the company is going to be real scared to hear of the big militant turnout at this strike meeting. From what I hear, this meeting is about eight times bigger than most local meetings, and Piasecki is probably just as nervous as the company right now.

He and other union officials give long, empty reports about the strike talks, and it's obvious they want to drag this thing out until we become confused and bored and hopefully we'll go home. After a half-hour some of the guys on the floor start to yell out. Piasecki takes a big sheaf of papers from his briefcase and shouts, "Brothers! I hope we are going to register a big strike-vote here tonight, to give our union leadership the backing they deserve." He raises the messy sheaf in the air. "You know the company is stalling and won't cooperate. And I'd bet you are wanting to know just how these negotiations stand."

"Yeah!" "Yeah!" "Tell us," the shouts ring out.

He begins. "I have a copy of our bargaining position right here. It has fourteen categories of demands on it, and the company has only agreed to talk about six of them. Even to talk! That's the kind of bad faith we're up against."

"What is it?" "What are they offering?" the cries go up.

"Right. Right," says the president. "And I'm going to tell you exactly what these categories are." He shuffles through papers some more, while we wait. Finally he says, loud and decisive, "Number three!" Then silence, and after more shuffling and a brief consultation with another official sitting at the microphone-covered table, "Number five! Number six!" Then more shuffling, and heads out in the audience are looking back and forth, and asses are shifting around in their chairs. Eventually he gets out the rest of the six numbers. Nothing more, just numbers. This is union democracy in action. Most of the audience is looking passive, but a few start to yell again.

"Hey, what's with these numbers, Harry?" "What are they offering?" "What's the old age benefit?" "Come on, what's going on?"

The gavel goes rap rap rap rap and Piasecki says, "O.K. brothers, let's keep it orderly, every man will have his chance to speak." But the uproar continues and finally he shouts, "Do you want to know what's in these categories?" Clearly they do. So he consults, shuffles more papers, and then begins to read out the answers. "Wages and Hours!" Pause. Shuffle. "Retirement Plans!" He goes on like this, with the category titles. That's all. Just the titles. Confusion and passivity increase. Some men yell. Some leave. Most just sit.

Now the microphone on the floor is opened up, but those who ask good questions don't get answers, and many of the speakers are minor union officials. They remind us that the union charter doesn't allow us to vote on the terms of the contract; it only allows us to authorize Piasecki to authorize Abel to call a strike. They talk drivel, but use tough, labor-movement language: "Brothers and sisters!!" "Standing bravely on the picket line!!" "We're not going to knuckle under to that goddamn company until Motherfucking Hell Freezes Over, excuse me if there are ladies present."

It's getting late, and more people leave.

Those who thought this meeting was actually going to mean something are frustrated and disappointed. At just the right point, Piasecki calls the vote. "All right men, how many are in favor of authorizing a strike against this company, if the negotiatiors feel it is necessary? Let's hear it!" There is a thunderous "Aye!" How could you vote anything else? Does anyone here want to vote in favor of the company? That's what he seems to ask as he shouts, "Those opposed?" We all look around, but no hands are raised. How could they be? This is union democracy.

Still, I can't feel too superior to the union when I look back over my own six months here. The direct actions I took part in were all for more bonus money or less work, and these are things the workers have already — with their job security and other past gains protected by the union.

This current set-up can pretty much handle the traditional issues. There's a need for changes here, but it's around a whole new type of things.

Just before the strike, Billy and I have a little talk about "life." To make some point, I ask about his boat.

He says, "When the strike comes, I may finally have time to paint her up and get her into the water." It's almost August and he hasn't used it yet. Billy gives me a sheepish look.

I think the deepest needs of my friends here, the things that require radical changes, are the same kind of unclear things that once made me want to wear long hair and raggy clothes, that made me rebel against the Vietnam war and ask the most basic questions about my own life. For a sense of revolution was in the air. I felt like history was ready for the development of a whole new kind of person. Things like art, communtiy, sex roles, civil rights, a different sense of work, participatory democracy, creativity — somehow things like this were being remolded into a new vision.

Then came the worst years of the war and the worst years of Nixon. Cynicism took over. I began to abandon the dreams that I cared most about. But when I look back over my time in the mill, it's clear that it's exactly that newer, freer, higher vision that people need. It's the one real thing that Billy and my other friends could really throw their lives into.

On July 31, the day before the strike is

supposed to begin, newspaper headlines trumpet: The union and the company have Reached an Agreement! At the same time the company announces a layoff of practically everybody in the plant — for three, four, possibly five months. I realize that I have just witnessed The Great Steel Strike. Or The Great Steel Layoff. It's a farce, or a ritual. What's truly happening isn't in the papers yet. It's still smoldering underneath the surface.

I walk around the mill saying goodbye to friends everywhere — Hester, Sammy, David, Chico, Tom-bone, Cook, and many more. There is such a mixture of hope, high spirits, and sadness. U. S. Steel. I steal my hard hat for a keepsake when I leave.

Stephen Packard was young in suburban Massachusetts, which he "tried to escape into the woods from, every day after school." Later he moved to various cities to be an artist, college student, dock worker, etc.

During the Vietnam disaster he was active in the anti-war movement and in the other political and social experimentation of the times, which had a deep impact on him. Now he lives in Chicago and works with the Illinois Environmental Council — an activist citizens' organization.